Preventative Maintenance &

Fill in your Manufacturers/Company Requireme

Keep this log in a safe place to reference for future service or Warranty Work.

Transfer this documented service & repair information to a new operator or owner.

Record any specific information about this equipment below.

MW01537278

PM Number _ _ _ _	HOURS_ _ _ _ _ _	DATE:_ _ _
REQUIRED OPERATION	PART NUMBERS	COST $
COMPLETED BY NAME:	SIGN:	

DATE	INITIALS	OPERATOR NOTES / OTHER SERVICE AND PARTS	COST $

NOTES	SAMPLES OR TESTING COMPLETED

PM Number _ _ _ _	HOURS _ _ _ _ _ _	DATE:_ _ _
REQUIRED OPERATION	PART NUMBERS	COST $
COMPLETED BY NAME:	SIGN:	

DATE	INITIALS	OPERATOR NOTES / OTHER SERVICE AND PARTS	COST $

NOTES	SAMPLES OR TESTING COMPLETED

PM Number _ _ _ _	HOURS _ _ _ _ _ _	DATE: _ _ _
REQUIRED OPERATION	PART NUMBERS	COST $
COMPLETED BY NAME:	SIGN:	

DATE	INITIALS	OPERATOR NOTES / OTHER SERVICE AND PARTS	COST $

NOTES	SAMPLES OR TESTING COMPLETED

PM Number _ _ _ _	HOURS_ _ _ _ _ _	DATE:_ _ _
REQUIRED OPERATION	PART NUMBERS	COST $
COMPLETED BY NAME:	SIGN:	

DATE	INITIALS	OPERATOR NOTES / OTHER SERVICE AND PARTS	COST $

NOTES	SAMPLES OR TESTING COMPLETED

PM Number _ _ _ _	HOURS _ _ _ _ _ _	DATE:_ _ _
REQUIRED OPERATION	PART NUMBERS	COST $
COMPLETED BY NAME:	SIGN:	

DATE	INITIALS	OPERATOR NOTES / OTHER SERVICE AND PARTS	COST $

NOTES	SAMPLES OR TESTING COMPLETED

PM Number _ _ _ _	HOURS _ _ _ _ _ _	DATE:_ _ _
REQUIRED OPERATION	PART NUMBERS	COST $
COMPLETED BY NAME:	SIGN:	

DATE	INITIALS	OPERATOR NOTES / OTHER SERVICE AND PARTS	COST $

NOTES	SAMPLES OR TESTING COMPLETED

PM Number _ _ _ _	HOURS _ _ _ _ _ _	DATE: _ _ _
REQUIRED OPERATION	PART NUMBERS	COST $
COMPLETED BY NAME:	SIGN:	

DATE	INITIALS	OPERATOR NOTES / OTHER SERVICE AND PARTS	COST $

NOTES	SAMPLES OR TESTING COMPLETED

PM Number _ _ _ _	HOURS _ _ _ _ _ _	DATE: _ _ _
REQUIRED OPERATION	PART NUMBERS	COST $
COMPLETED BY NAME:	SIGN:	

DATE	INITIALS	OPERATOR NOTES / OTHER SERVICE AND PARTS	COST $

NOTES	SAMPLES OR TESTING COMPLETED

PM Number _ _ _ _	HOURS _ _ _ _ _ _	DATE:_ _ _
REQUIRED OPERATION	PART NUMBERS	COST $
COMPLETED BY NAME:	SIGN:	

DATE	INITIALS	OPERATOR NOTES / OTHER SERVICE AND PARTS	COST $

NOTES	SAMPLES OR TESTING COMPLETED

PM Number _ _ _ _	HOURS _ _ _ _ _ _	DATE:_ _ _
REQUIRED OPERATION	PART NUMBERS	COST $
COMPLETED BY NAME:	SIGN:	

DATE	INITIALS	OPERATOR NOTES / OTHER SERVICE AND PARTS	COST $

NOTES	SAMPLES OR TESTING COMPLETED

PM Number _ _ _ _	HOURS _ _ _ _ _ _	DATE: _ _ _
REQUIRED OPERATION	PART NUMBERS	COST $
COMPLETED BY NAME:	SIGN:	

DATE	INITIALS	OPERATOR NOTES / OTHER SERVICE AND PARTS	COST $

NOTES	SAMPLES OR TESTING COMPLETED

PM Number _ _ _ _	HOURS _ _ _ _ _ _	DATE:_ _ _
REQUIRED OPERATION	PART NUMBERS	COST $
COMPLETED BY NAME:	SIGN:	

DATE	INITIALS	OPERATOR NOTES / OTHER SERVICE AND PARTS	COST $

NOTES	SAMPLES OR TESTING COMPLETED

PM Number _ _ _ _	HOURS _ _ _ _ _ _	DATE:_ _ _
REQUIRED OPERATION	PART NUMBERS	COST $
COMPLETED BY NAME:	SIGN:	

DATE	INITIALS	OPERATOR NOTES / OTHER SERVICE AND PARTS	COST $

NOTES	SAMPLES OR TESTING COMPLETED

PM Number _ _ _ _	HOURS _ _ _ _ _ _	DATE:_ _ _
REQUIRED OPERATION	PART NUMBERS	COST $
COMPLETED BY NAME:	SIGN:	

DATE	INITIALS	OPERATOR NOTES / OTHER SERVICE AND PARTS	COST $

NOTES	SAMPLES OR TESTING COMPLETED

PM Number _ _ _ _	HOURS _ _ _ _ _ _	DATE: _ _ _
REQUIRED OPERATION	**PART NUMBERS**	**COST $**
COMPLETED BY NAME:	**SIGN:**	

DATE	INITIALS	OPERATOR NOTES / OTHER SERVICE AND PARTS	COST $

NOTES	SAMPLES OR TESTING COMPLETED

PM Number _ _ _ _	HOURS _ _ _ _ _ _	DATE: _ _ _
REQUIRED OPERATION	PART NUMBERS	COST $
COMPLETED BY NAME:	SIGN:	

DATE	INITIALS	OPERATOR NOTES / OTHER SERVICE AND PARTS	COST $

NOTES	SAMPLES OR TESTING COMPLETED

PM Number _ _ _ _	HOURS _ _ _ _ _ _	DATE: _ _ _
REQUIRED OPERATION	PART NUMBERS	COST $
COMPLETED BY NAME:	SIGN:	

DATE	INITIALS	OPERATOR NOTES / OTHER SERVICE AND PARTS	COST $

NOTES	SAMPLES OR TESTING COMPLETED

PM Number _ _ _ _	HOURS_ _ _ _ _ _	DATE:_ _ _
REQUIRED OPERATION	PART NUMBERS	COST $
COMPLETED BY NAME:	SIGN:	

DATE	INITIALS	OPERATOR NOTES / OTHER SERVICE AND PARTS	COST $

NOTES	SAMPLES OR TESTING COMPLETED

PM Number _ _ _ _	HOURS _ _ _ _ _ _	DATE: _ _ _
REQUIRED OPERATION	PART NUMBERS	COST $
COMPLETED BY NAME:	SIGN:	

DATE	INITIALS	OPERATOR NOTES / OTHER SERVICE AND PARTS	COST $

NOTES	SAMPLES OR TESTING COMPLETED

PM Number _ _ _ _	HOURS _ _ _ _ _ _	DATE: _ _ _
REQUIRED OPERATION	PART NUMBERS	COST $
COMPLETED BY NAME:	SIGN:	

DATE	INITIALS	OPERATOR NOTES / OTHER SERVICE AND PARTS	COST $

NOTES	SAMPLES OR TESTING COMPLETED

PM Number _ _ _ _	HOURS _ _ _ _ _ _	DATE: _ _ _
REQUIRED OPERATION	PART NUMBERS	COST $
COMPLETED BY NAME:	SIGN:	

DATE	INITIALS	OPERATOR NOTES / OTHER SERVICE AND PARTS	COST $

NOTES	SAMPLES OR TESTING COMPLETED

PM Number _ _ _ _	HOURS _ _ _ _ _ _	DATE: _ _ _
REQUIRED OPERATION	PART NUMBERS	COST $
COMPLETED BY NAME:	SIGN:	

DATE	INITIALS	OPERATOR NOTES / OTHER SERVICE AND PARTS	COST $

NOTES	SAMPLES OR TESTING COMPLETED

PM Number _ _ _ _	HOURS _ _ _ _ _ _	DATE: _ _ _
REQUIRED OPERATION	PART NUMBERS	COST $
COMPLETED BY NAME:	SIGN:	

DATE	INITIALS	OPERATOR NOTES / OTHER SERVICE AND PARTS	COST $

NOTES	SAMPLES OR TESTING COMPLETED

PM Number _ _ _ _	HOURS _ _ _ _ _ _	DATE: _ _ _
REQUIRED OPERATION	PART NUMBERS	COST $
COMPLETED BY NAME:	SIGN:	

DATE	INITIALS	OPERATOR NOTES / OTHER SERVICE AND PARTS	COST $

NOTES	SAMPLES OR TESTING COMPLETED

PM Number _ _ _ _	HOURS _ _ _ _ _ _	DATE: _ _ _
REQUIRED OPERATION	PART NUMBERS	COST $
COMPLETED BY NAME:	SIGN:	

DATE	INITIALS	OPERATOR NOTES / OTHER SERVICE AND PARTS	COST $

NOTES	SAMPLES OR TESTING COMPLETED

PM Number _ _ _ _	HOURS _ _ _ _ _ _	DATE:_ _ _
REQUIRED OPERATION	PART NUMBERS	COST $
COMPLETED BY NAME:	SIGN:	

DATE	INITIALS	OPERATOR NOTES / OTHER SERVICE AND PARTS	COST $

NOTES	SAMPLES OR TESTING COMPLETED

PM Number _ _ _ _	HOURS _ _ _ _ _ _	DATE: _ _ _
REQUIRED OPERATION	PART NUMBERS	COST $
COMPLETED BY NAME:	SIGN:	

DATE	INITIALS	OPERATOR NOTES / OTHER SERVICE AND PARTS	COST $

NOTES	SAMPLES OR TESTING COMPLETED

PM Number _ _ _ _	HOURS _ _ _ _ _ _	DATE: _ _ _
REQUIRED OPERATION	PART NUMBERS	COST $
COMPLETED BY NAME:	SIGN:	

DATE	INITIALS	OPERATOR NOTES / OTHER SERVICE AND PARTS	COST $

NOTES	SAMPLES OR TESTING COMPLETED

PM Number _ _ _ _	HOURS _ _ _ _ _ _ _	DATE: _ _ _
REQUIRED OPERATION	PART NUMBERS	COST $
COMPLETED BY NAME:	SIGN:	

DATE	INITIALS	OPERATOR NOTES / OTHER SERVICE AND PARTS	COST $

NOTES	SAMPLES OR TESTING COMPLETED

PM Number _ _ _ _	HOURS _ _ _ _ _ _	DATE:_ _ _
REQUIRED OPERATION	PART NUMBERS	COST $
COMPLETED BY NAME:	SIGN:	

DATE	INITIALS	OPERATOR NOTES / OTHER SERVICE AND PARTS	COST $

NOTES	SAMPLES OR TESTING COMPLETED

PM Number _ _ _ _	HOURS _ _ _ _ _ _	DATE: _ _ _
REQUIRED OPERATION	PART NUMBERS	COST $
COMPLETED BY NAME:	SIGN:	

DATE	INITIALS	OPERATOR NOTES / OTHER SERVICE AND PARTS	COST $

NOTES	SAMPLES OR TESTING COMPLETED

PM Number _ _ _ _	HOURS _ _ _ _ _ _	DATE:_ _ _
REQUIRED OPERATION	PART NUMBERS	COST $
COMPLETED BY NAME:	SIGN:	

DATE	INITIALS	OPERATOR NOTES / OTHER SERVICE AND PARTS	COST $

NOTES	SAMPLES OR TESTING COMPLETED

PM Number _ _ _ _	HOURS _ _ _ _ _ _	DATE: _ _ _
REQUIRED OPERATION	PART NUMBERS	COST $
COMPLETED BY NAME:	SIGN:	

DATE	INITIALS	OPERATOR NOTES / OTHER SERVICE AND PARTS	COST $

NOTES	SAMPLES OR TESTING COMPLETED

PM Number _ _ _ _	HOURS_ _ _ _ _ _	DATE:_ _ _
REQUIRED OPERATION	PART NUMBERS	COST $
COMPLETED BY NAME:	SIGN:	

DATE	INITIALS	OPERATOR NOTES / OTHER SERVICE AND PARTS	COST $

NOTES	SAMPLES OR TESTING COMPLETED

PM Number _ _ _ _	HOURS _ _ _ _ _ _	DATE:_ _ _
REQUIRED OPERATION	PART NUMBERS	COST $
COMPLETED BY NAME:	SIGN:	

DATE	INITIALS	OPERATOR NOTES / OTHER SERVICE AND PARTS	COST $

NOTES	SAMPLES OR TESTING COMPLETED

PM Number _ _ _ _	HOURS _ _ _ _ _ _	DATE: _ _ _
REQUIRED OPERATION	PART NUMBERS	COST $
COMPLETED BY NAME:	SIGN:	

DATE	INITIALS	OPERATOR NOTES / OTHER SERVICE AND PARTS	COST $

NOTES	SAMPLES OR TESTING COMPLETED

PM Number _ _ _ _	HOURS _ _ _ _ _ _	DATE:_ _ _
REQUIRED OPERATION	PART NUMBERS	COST $
COMPLETED BY NAME:	SIGN:	

DATE	INITIALS	OPERATOR NOTES / OTHER SERVICE AND PARTS	COST $

NOTES	SAMPLES OR TESTING COMPLETED

PM Number _ _ _ _	HOURS _ _ _ _ _ _	DATE: _ _ _
REQUIRED OPERATION	PART NUMBERS	COST $
COMPLETED BY NAME:	SIGN:	

DATE	INITIALS	OPERATOR NOTES / OTHER SERVICE AND PARTS	COST $

NOTES	SAMPLES OR TESTING COMPLETED

PM Number _ _ _ _	HOURS _ _ _ _ _ _	DATE: _ _ _
REQUIRED OPERATION	PART NUMBERS	COST $
COMPLETED BY NAME:	SIGN:	

DATE	INITIALS	OPERATOR NOTES / OTHER SERVICE AND PARTS	COST $

NOTES	SAMPLES OR TESTING COMPLETED

PM Number _ _ _ _	HOURS _ _ _ _ _ _	DATE:_ _ _
REQUIRED OPERATION	PART NUMBERS	COST $
COMPLETED BY NAME:	SIGN:	

DATE	INITIALS	OPERATOR NOTES / OTHER SERVICE AND PARTS	COST $

NOTES	SAMPLES OR TESTING COMPLETED

PM Number _ _ _ _	HOURS _ _ _ _ _ _	DATE:_ _ _
REQUIRED OPERATION	PART NUMBERS	COST $
COMPLETED BY NAME:	SIGN:	

DATE	INITIALS	OPERATOR NOTES / OTHER SERVICE AND PARTS	COST $

NOTES	SAMPLES OR TESTING COMPLETED

PM Number _ _ _ _	HOURS _ _ _ _ _ _	DATE: _ _ _
REQUIRED OPERATION	PART NUMBERS	COST $
COMPLETED BY NAME:	SIGN:	

DATE	INITIALS	OPERATOR NOTES / OTHER SERVICE AND PARTS	COST $

NOTES	SAMPLES OR TESTING COMPLETED

PM Number _ _ _ _	HOURS _ _ _ _ _ _	DATE: _ _ _
REQUIRED OPERATION	PART NUMBERS	COST $
COMPLETED BY NAME:	SIGN:	

DATE	INITIALS	OPERATOR NOTES / OTHER SERVICE AND PARTS	COST $

NOTES	SAMPLES OR TESTING COMPLETED

PM Number _ _ _ _	HOURS _ _ _ _ _ _	DATE: _ _ _
REQUIRED OPERATION	PART NUMBERS	COST $
COMPLETED BY NAME:	SIGN:	

DATE	INITIALS	OPERATOR NOTES / OTHER SERVICE AND PARTS	COST $

NOTES	SAMPLES OR TESTING COMPLETED

PM Number _ _ _ _	HOURS _ _ _ _ _ _	DATE: _ _ _
REQUIRED OPERATION	PART NUMBERS	COST $
COMPLETED BY NAME:	SIGN:	

DATE	INITIALS	OPERATOR NOTES / OTHER SERVICE AND PARTS	COST $

NOTES	SAMPLES OR TESTING COMPLETED

PM Number _ _ _ _	HOURS _ _ _ _ _ _	DATE: _ _ _
REQUIRED OPERATION	PART NUMBERS	COST $
COMPLETED BY NAME:	SIGN:	

DATE	INITIALS	OPERATOR NOTES / OTHER SERVICE AND PARTS	COST $

NOTES	SAMPLES OR TESTING COMPLETED

PM Number _ _ _ _	HOURS _ _ _ _ _ _	DATE: _ _ _
REQUIRED OPERATION	PART NUMBERS	COST $
COMPLETED BY NAME:	SIGN:	

DATE	INITIALS	OPERATOR NOTES / OTHER SERVICE AND PARTS	COST $

NOTES	SAMPLES OR TESTING COMPLETED

PM Number _ _ _ _	HOURS _ _ _ _ _ _	DATE:_ _ _
REQUIRED OPERATION	PART NUMBERS	COST $
COMPLETED BY NAME:	SIGN:	

DATE	INITIALS	OPERATOR NOTES / OTHER SERVICE AND PARTS	COST $

NOTES	SAMPLES OR TESTING COMPLETED

PM Number _ _ _ _	HOURS _ _ _ _ _ _	DATE:_ _ _
REQUIRED OPERATION	PART NUMBERS	COST $
COMPLETED BY NAME:	SIGN:	

DATE	INITIALS	OPERATOR NOTES / OTHER SERVICE AND PARTS	COST $

NOTES	SAMPLES OR TESTING COMPLETED

PM Number _ _ _ _	HOURS _ _ _ _ _ _	DATE:_ _ _
REQUIRED OPERATION	PART NUMBERS	COST $
COMPLETED BY NAME:	SIGN:	

DATE	INITIALS	OPERATOR NOTES / OTHER SERVICE AND PARTS	COST $

NOTES	SAMPLES OR TESTING COMPLETED

PM Number _ _ _ _	HOURS _ _ _ _ _ _	DATE:_ _ _
REQUIRED OPERATION	PART NUMBERS	COST $
COMPLETED BY NAME:	SIGN:	

DATE	INITIALS	OPERATOR NOTES / OTHER SERVICE AND PARTS	COST $

NOTES	SAMPLES OR TESTING COMPLETED

PM Number _ _ _ _	HOURS _ _ _ _ _ _	DATE: _ _ _
REQUIRED OPERATION	PART NUMBERS	COST $
COMPLETED BY NAME:	SIGN:	

DATE	INITIALS	OPERATOR NOTES / OTHER SERVICE AND PARTS	COST $

NOTES	SAMPLES OR TESTING COMPLETED

PM Number _ _ _ _	HOURS _ _ _ _ _ _	DATE: _ _ _
REQUIRED OPERATION	PART NUMBERS	COST $
COMPLETED BY NAME:	SIGN:	

DATE	INITIALS	OPERATOR NOTES / OTHER SERVICE AND PARTS	COST $

NOTES	SAMPLES OR TESTING COMPLETED

PM Number _ _ _ _	HOURS _ _ _ _ _ _	DATE:_ _ _
REQUIRED OPERATION	PART NUMBERS	COST $
COMPLETED BY NAME:	SIGN:	

DATE	INITIALS	OPERATOR NOTES / OTHER SERVICE AND PARTS	COST $

NOTES	SAMPLES OR TESTING COMPLETED

PM Number _ _ _ _	HOURS _ _ _ _ _ _	DATE:_ _ _
REQUIRED OPERATION	PART NUMBERS	COST $
COMPLETED BY NAME:	SIGN:	

DATE	INITIALS	OPERATOR NOTES / OTHER SERVICE AND PARTS	COST $

NOTES	SAMPLES OR TESTING COMPLETED

PM Number _ _ _ _	HOURS _ _ _ _ _ _	DATE: _ _ _
REQUIRED OPERATION	PART NUMBERS	COST $
COMPLETED BY NAME:	SIGN:	

DATE	INITIALS	OPERATOR NOTES / OTHER SERVICE AND PARTS	COST $

NOTES	SAMPLES OR TESTING COMPLETED

PM Number _ _ _ _	HOURS _ _ _ _ _ _	DATE:_ _ _
REQUIRED OPERATION	PART NUMBERS	COST $
COMPLETED BY NAME:	SIGN:	

DATE	INITIALS	OPERATOR NOTES / OTHER SERVICE AND PARTS	COST $

NOTES	SAMPLES OR TESTING COMPLETED

PM Number _ _ _ _	HOURS _ _ _ _ _ _	DATE: _ _ _
REQUIRED OPERATION	PART NUMBERS	COST $
COMPLETED BY NAME:	SIGN:	

DATE	INITIALS	OPERATOR NOTES / OTHER SERVICE AND PARTS	COST $

NOTES	SAMPLES OR TESTING COMPLETED

PM Number _ _ _ _	HOURS _ _ _ _ _ _	DATE:_ _ _
REQUIRED OPERATION	PART NUMBERS	COST $
COMPLETED BY NAME:	SIGN:	

DATE	INITIALS	OPERATOR NOTES / OTHER SERVICE AND PARTS	COST $

NOTES	SAMPLES OR TESTING COMPLETED

Made in the USA
Las Vegas, NV
30 January 2022